EXPLORE THE WORLD

PHYSICAL SCIENCE

Artificial Body Parts

DEBBIE VILARDI

TABLE OF CONTENTS

Fulfilling a Need 2
Creating a Kidney 4
A Man-Made Heart 7
Artificial Limbs throughout History 10
Printing Parts 14
Pondering Problems and
Looking Forward 16
Glossary/Index 20

PIONEER VALLEY EDUCATIONAL PRESS, INC

FULFILLING A NEED

Imagine you're at a dinner party with President George Washington and a ship captain who helped the United States defeat the British during the American Revolution. The captain hobbles toward you on a peg leg and extends his right hand. As you shake hands, you notice his other arm hangs at his side. A hook has replaced his left hand.

You all sit down for dinner, and the president adjusts his false teeth. As the president tries to get comfortable, the captain explains how he received his injuries. You wonder about the lives of these fascinating men.

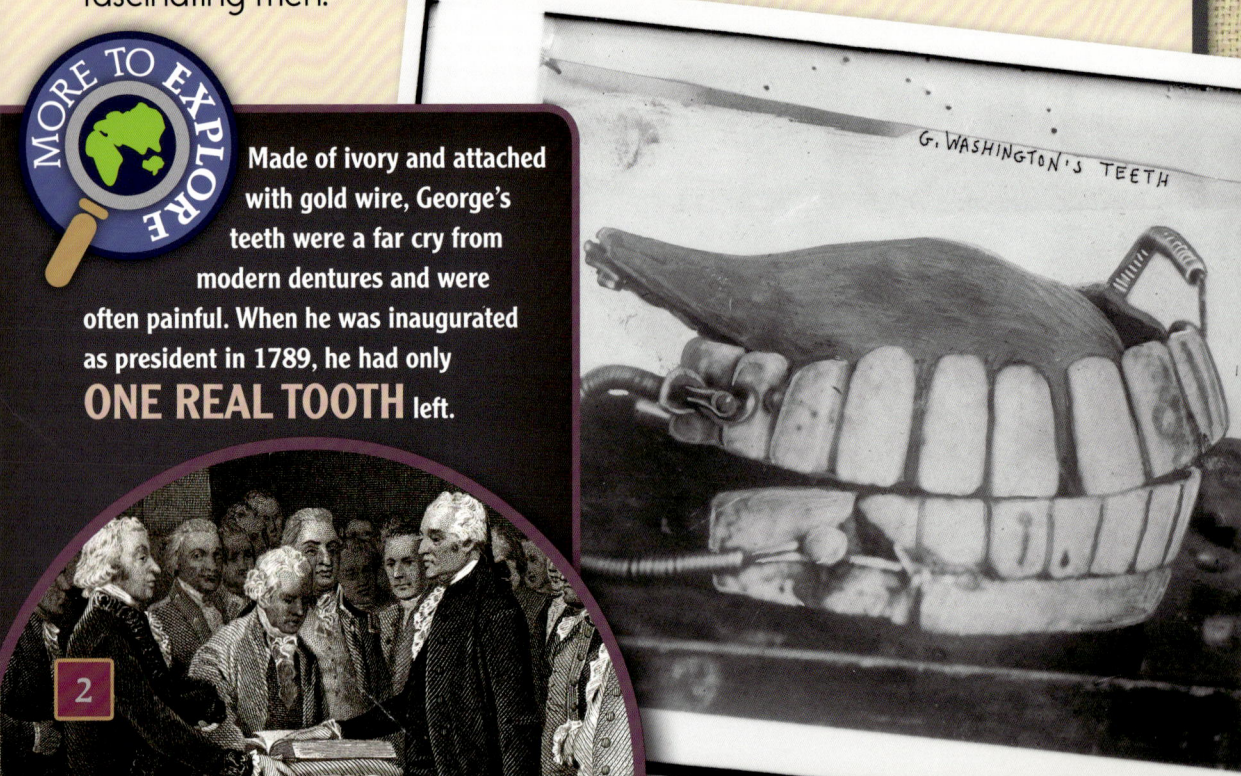

MORE TO EXPLORE

Made of ivory and attached with gold wire, George's teeth were a far cry from modern dentures and were often painful. When he was inaugurated as president in 1789, he had only **ONE REAL TOOTH** left.

The first known use of **prostheses** occurred in Egypt almost 3,000 years ago. Archaeologists discovered the remains of a noblewoman wearing a wooden toe that looked very much like a real **digit**. Researchers believe that the substitute toe may have helped her with balance.

From their earliest existence, the primary purpose of many prostheses has been to help their wearers fit in and function independently. Some prostheses help people accomplish everyday tasks that others take for granted, such as taking a walk or brushing their teeth. Others do even more; they save lives.

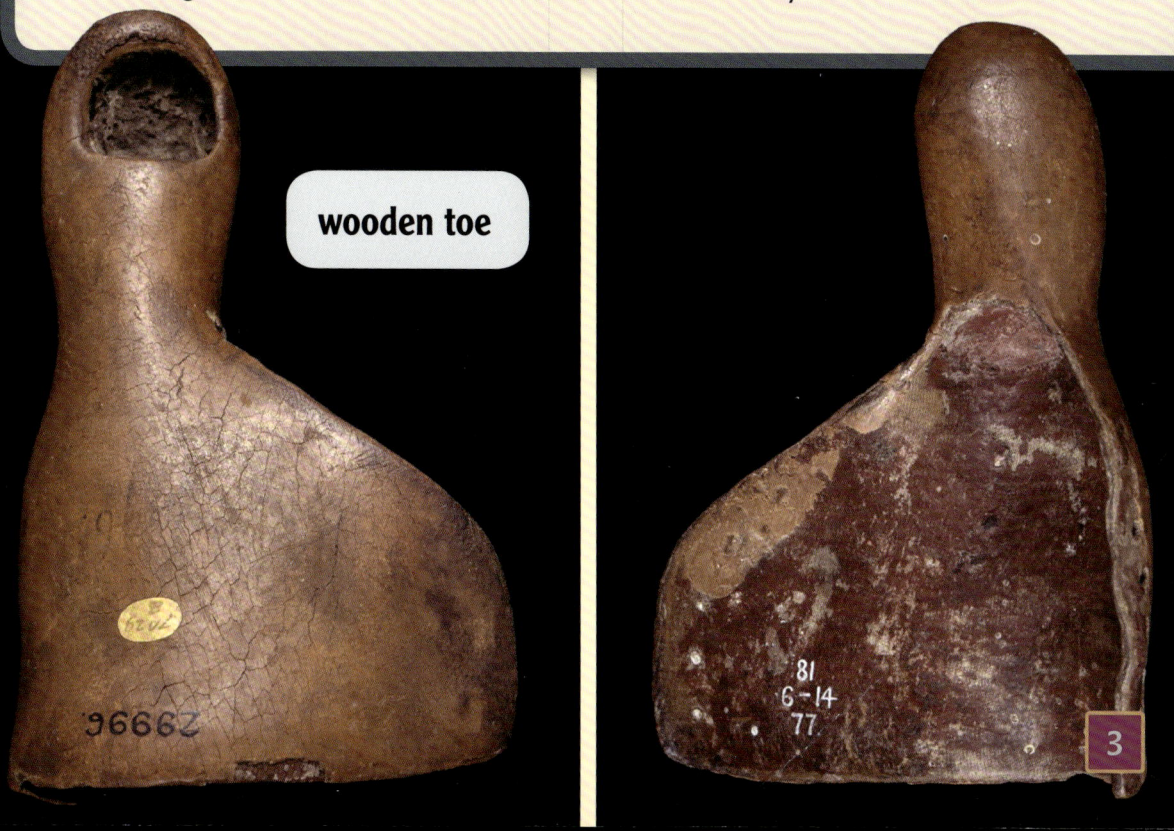

wooden toe

CREATING A KIDNEY

The kidneys are the body's filtration system. Like a filter in a swimming pool, they remove what doesn't belong. If kidneys don't function properly, excess fluid and toxins build up in the bloodstream. This leads to health problems that could be fatal. When a transplant is not possible, artificial kidneys help clean the blood.

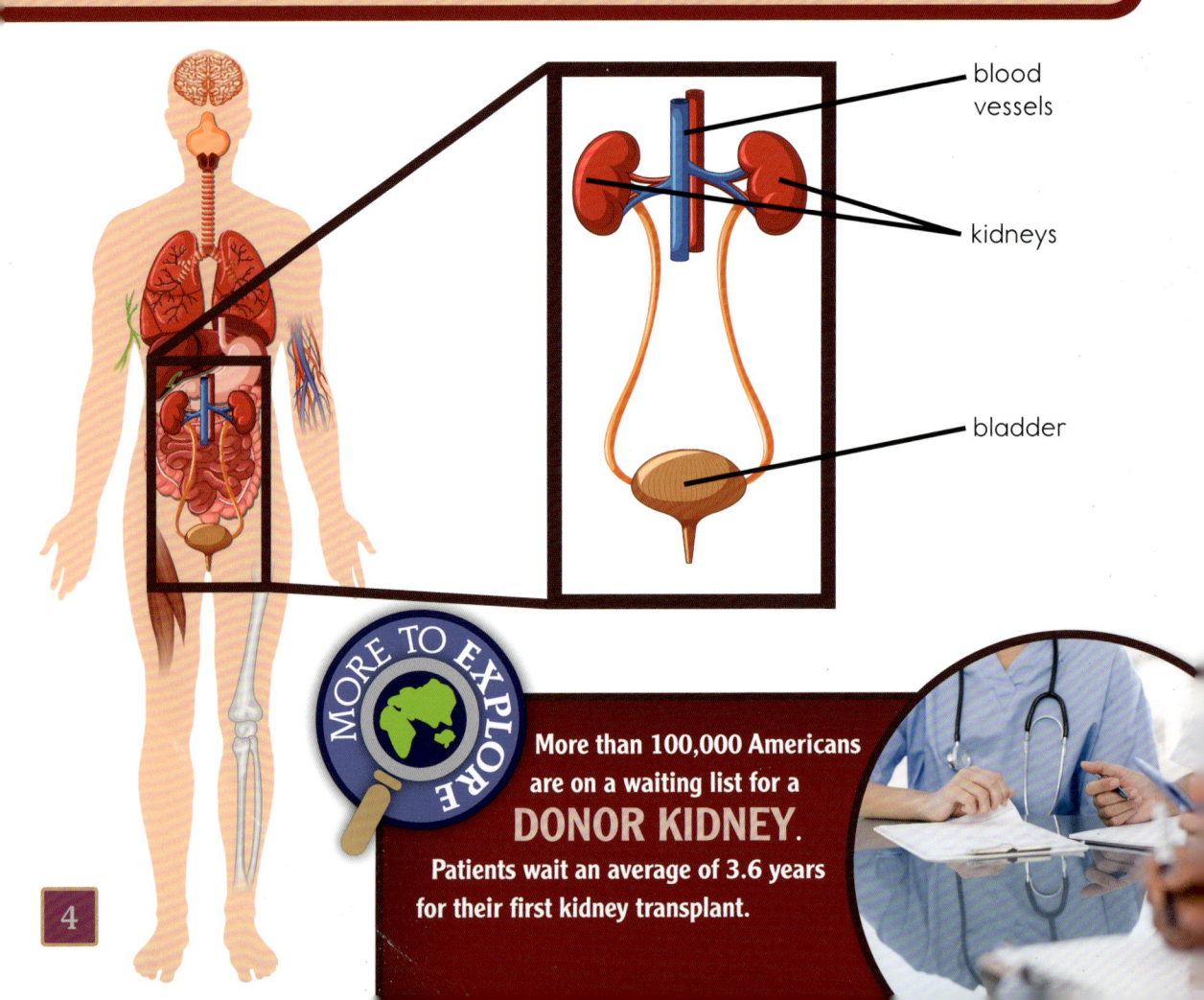

- blood vessels
- kidneys
- bladder

MORE TO EXPLORE

More than 100,000 Americans are on a waiting list for a **DONOR KIDNEY**. Patients wait an average of 3.6 years for their first kidney transplant.

Built in 1943, the first artificial kidneys were dialysis machines, which sit outside the body and filter the blood when the kidneys are not sufficient. Patients may receive treatment three times a week at a dialysis center or daily at home. The invention of dialysis turned kidney disease from a deadly condition into a treatable **chronic** illness.

Dialysis has saved many lives, but having to visit a center or hospital frequently can become a burden. Also, only about one in three patients on dialysis survives for longer than five years without an **organ** transplant.

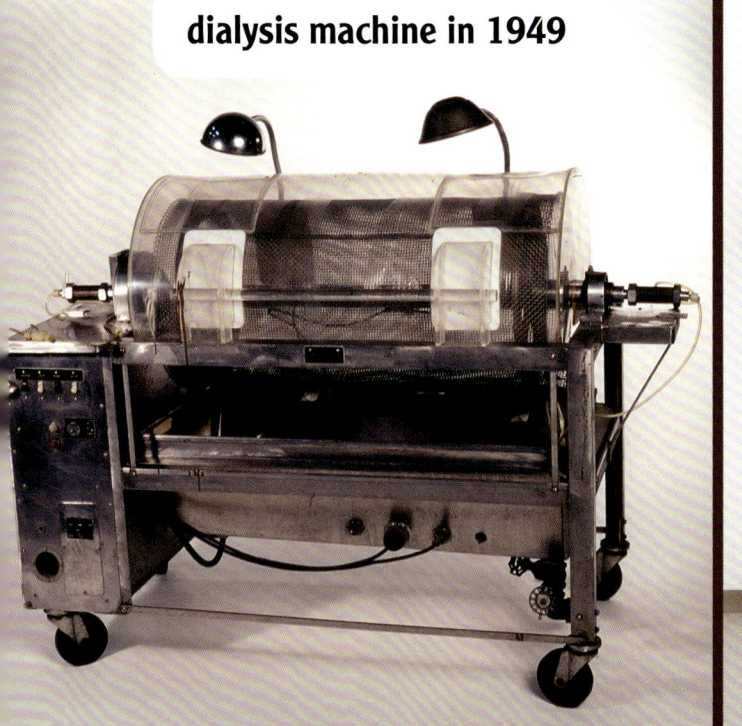

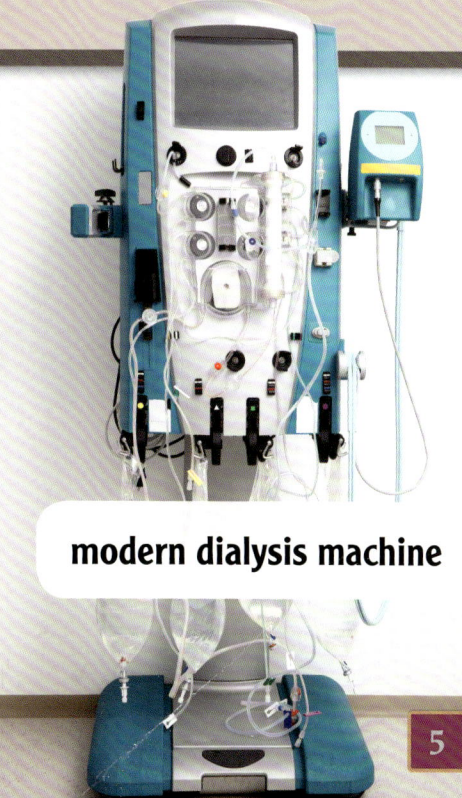

dialysis machine in 1949

modern dialysis machine

This has led scientists at The Kidney Project to work on another potential solution. They're developing an implantable bioartificial kidney that would use live kidney cells and microchip filters operated by the patient's heart to get rid of waste. It may have another benefit as well. Because this organ would use live cells, scientists believe there is little risk of the body **rejecting** it. They hope to start testing the device soon.

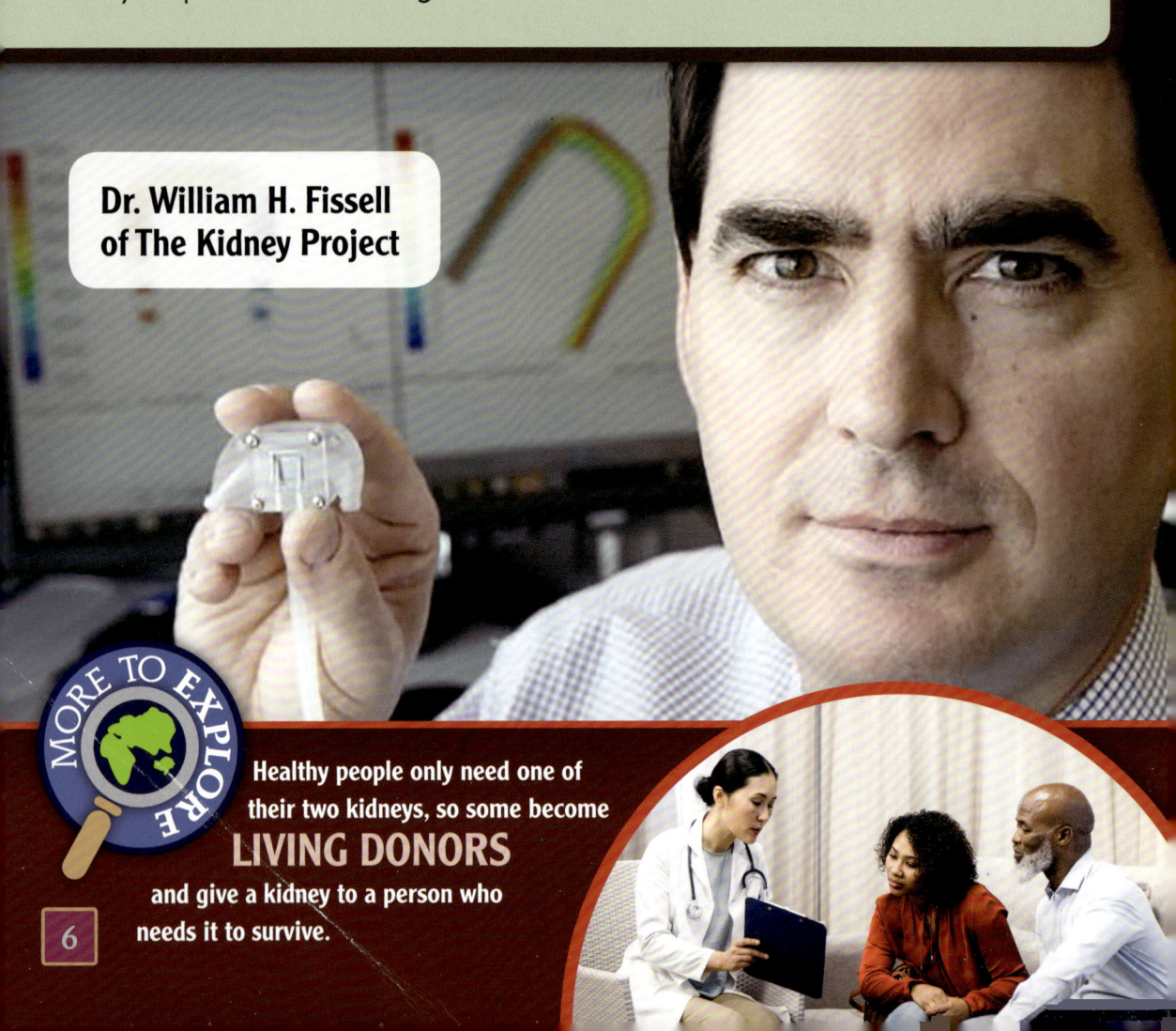

Dr. William H. Fissell of The Kidney Project

MORE TO EXPLORE

Healthy people only need one of their two kidneys, so some become **LIVING DONORS** and give a kidney to a person who needs it to survive.

A MAN-MADE HEART

A commonly used saying states that some people have a "heart of gold," meaning they are very kind. However, scientists and engineers have been working on building hearts out of metal and plastic.

It took a long time and many failed experiments before the first heart-lung machine made heart surgeries easier. The machine consists of tubes and a pump that **bypass** the patient's organs, so blood flow won't interfere with surgery. It does the job of both the heart and lungs.

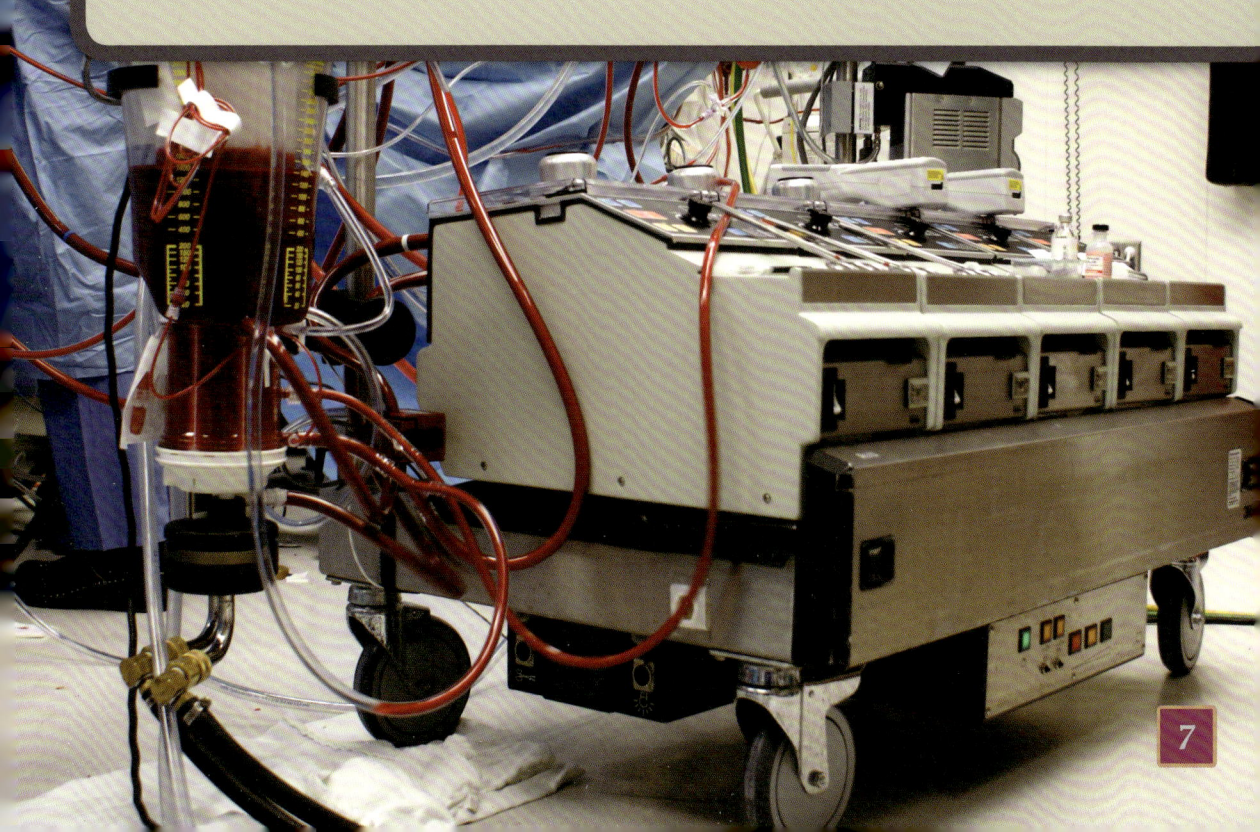

The heart-lung machine was a revolutionary device, but for some it wasn't the end goal. A total artificial heart (TAH) could save patients whose hearts weren't reparable. Again, there were many failures, but doctors persisted until they had a **viable** model.

In 1982, Dr. Barney Clark received the first Jarvik 7 artificial heart. He survived in the hospital for 112 days. Because the Jarvik 7 required external equipment to keep it running, Barney was hospitalized for the entire time. Today, scientists have a portable TAH, which is used by patients waiting for a transplant. It isn't meant to be a lifelong solution, but a permanent artificial heart may be on the horizon.

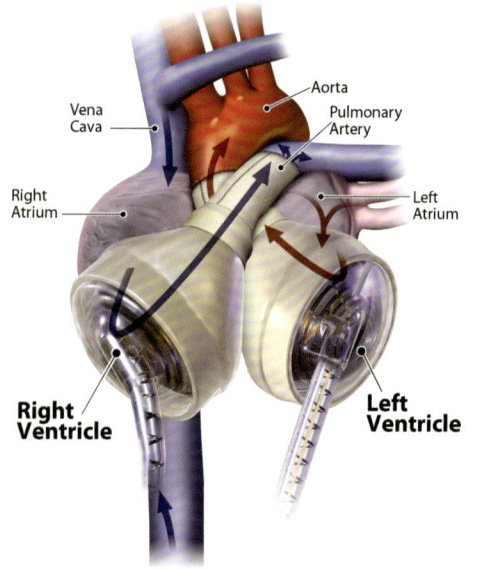

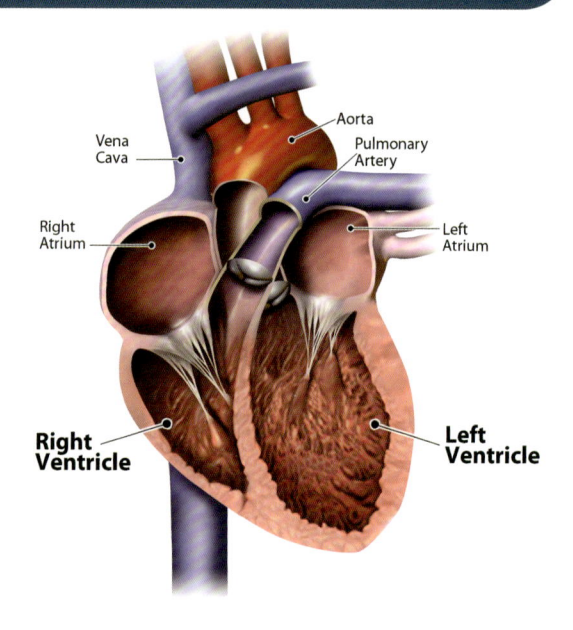

Total Artificial Heart

Human Heart

Not everyone with **cardiac** problems needs an entirely new heart. Some patients may have leaky valves replaced with artificial ones in order to maintain proper blood flow. People could have pacemakers implanted under the skin to help regulate their heartbeat. Others may get ventricular assist devices to help the ventricle, a part of the heart, pump blood to the rest of the body.

a pacemaker

MORE TO EXPLORE

The GREEK WORD KARDIA means heart. It is the origin of the term *cardiology*, which is the study of the heart.

ARTIFICIAL LIMBS THROUGHOUT HISTORY

In ancient times, artificial limbs were made of wood and metal. It took several people to develop the part: one might shape the wood as another addressed the metal. If the limb had moving parts, a watchmaker might design the inner workings. Later, leather and paper replaced wood and metal because they weigh less.

From AD 500 to 1000, a period known as the Dark Ages, artificial body parts were often given to people who were injured in battle. The peg leg and pirate's hook are a few familiar examples that were regularly used.

bronze leg, c. 300

iron arm, c. 1500

Replacement limbs continued to become more natural looking and functional until the time of the American Civil War. However, they didn't change or advance much between that time and World War II. Veterans of that conflict pushed the government to produce more modern prostheses.

nonlocking leg prosthesis, c. 1756

reconstructive plastic surgery, c. 1914

War injuries have been the driving force behind artificial limb production, since they are a common cause of amputations. Sometimes prostheses allow soldiers to return to battle. Other times, limbs with interchangeable ends enable the wearer to accomplish specific tasks, such as welding or eating. This promotes an individual's self-esteem by returning a sense of normalcy to their daily activities.

Today's prostheses contain internal computers and motors that allow them to function more like natural limbs. Many of these also look like the actual body part. This reduces the **stigma** of disability for the person wearing it.

MORE TO EXPLORE

Athletes use prostheses specifically designed for their sport. There are artificial limbs designed for **RUNNING**, swimming, motocross, baseball, hockey, rock climbing, and more.

PRINTING PARTS

Additive manufacturing, also called 3-D printing, is the creation of an object made by adding individual layers of a material, one layer at a time. It is at the cutting edge of the science of prosthetics.

How are scientists using 3-D printing? One innovation is the creation of artificial bones and teeth. Scientists are making teeth from a special plastic that fights cavities without the help of toothpaste. These chompers are still being tested and have a long way to go before they hit the market for humans.

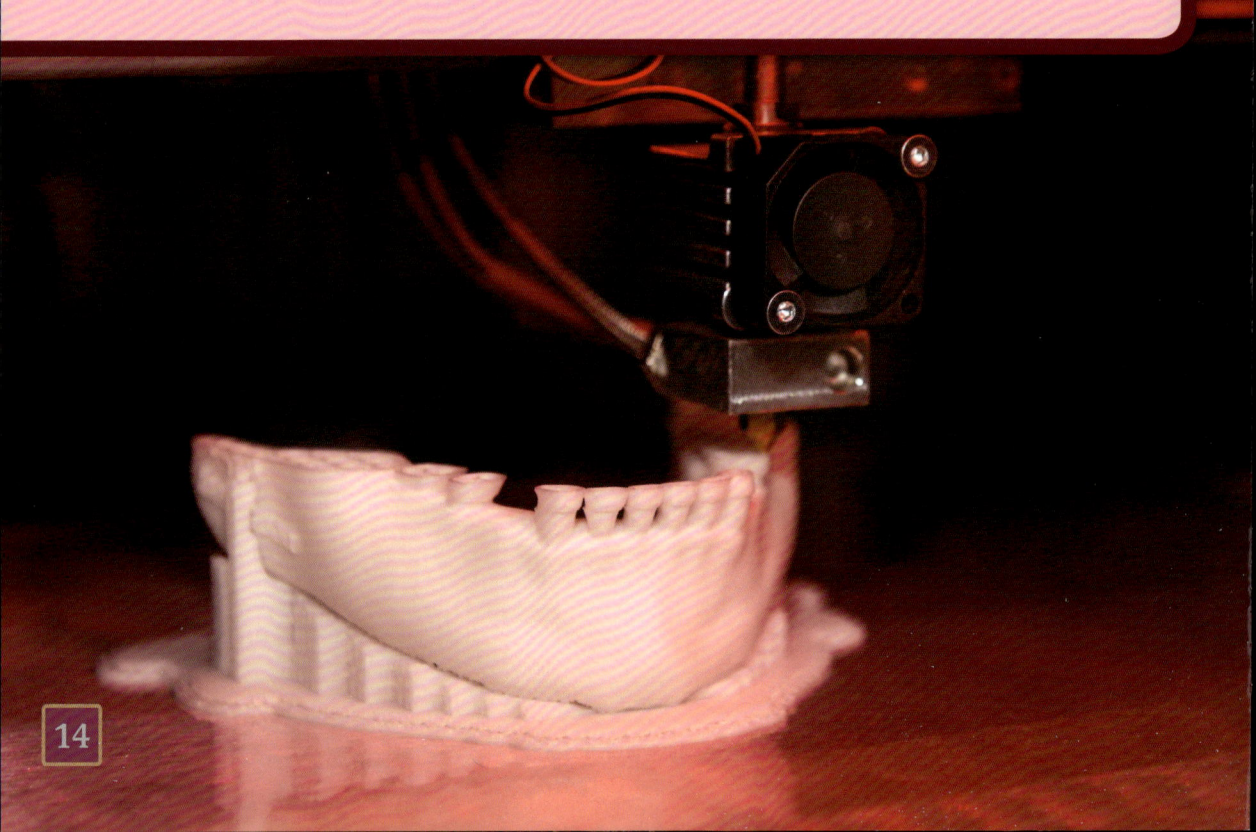

Another invention still under development is 3-D artificial blood vessels. The body's organs need blood vessels to deliver nutrients and remove waste products. Scientists are working on artificial blood vessels to supply both natural and artificial organs with blood. They coat the blood vessels with living cells so the body won't reject them. This advancement is also awaiting human **trials**.

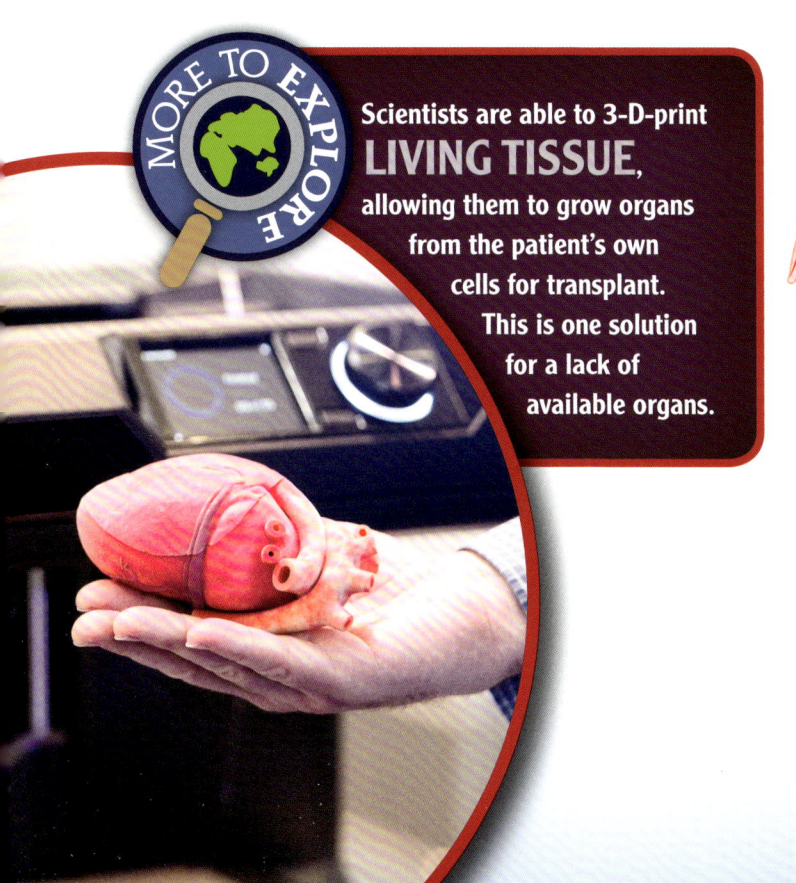

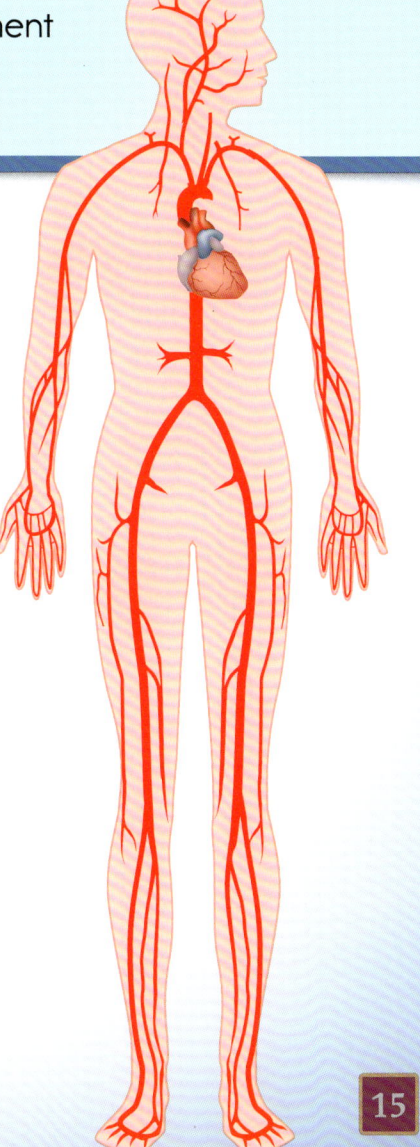

MORE TO EXPLORE

Scientists are able to 3-D-print **LIVING TISSUE**, allowing them to grow organs from the patient's own cells for transplant. This is one solution for a lack of available organs.

PONDERING PROBLEMS AND LOOKING FORWARD

Prostheses are designed to solve issues caused by disease and injuries, but they can have drawbacks.

One issue with some artificial devices is that they are linked to computers, so people fear they could be vulnerable to hacking. A hacked TAH might malfunction, endangering its user's life. Alternately, a person with an artificial arm might lose control of that limb if it was **commandeered** by a hacker. More work needs to be done to secure the safety of these connected devices.

Another challenge involves how prostheses communicate with the rest of the body. Although artificial body parts can operate in many of the same ways as what they replace, most do not function as sensory organs. So even though an artificial limb can move like a human one, it cannot transmit the sense of touch to the brain. That means the user does not feel heat, cold, or pain in that limb.

For scientists, the greatest success will happen when the brain and artificial device communicate naturally. Prosthetic hands with touch sensors that mimic nerves are in the works. Once the brain can receive sensory input from the prosthesis and respond to it, the user will have a full experience.

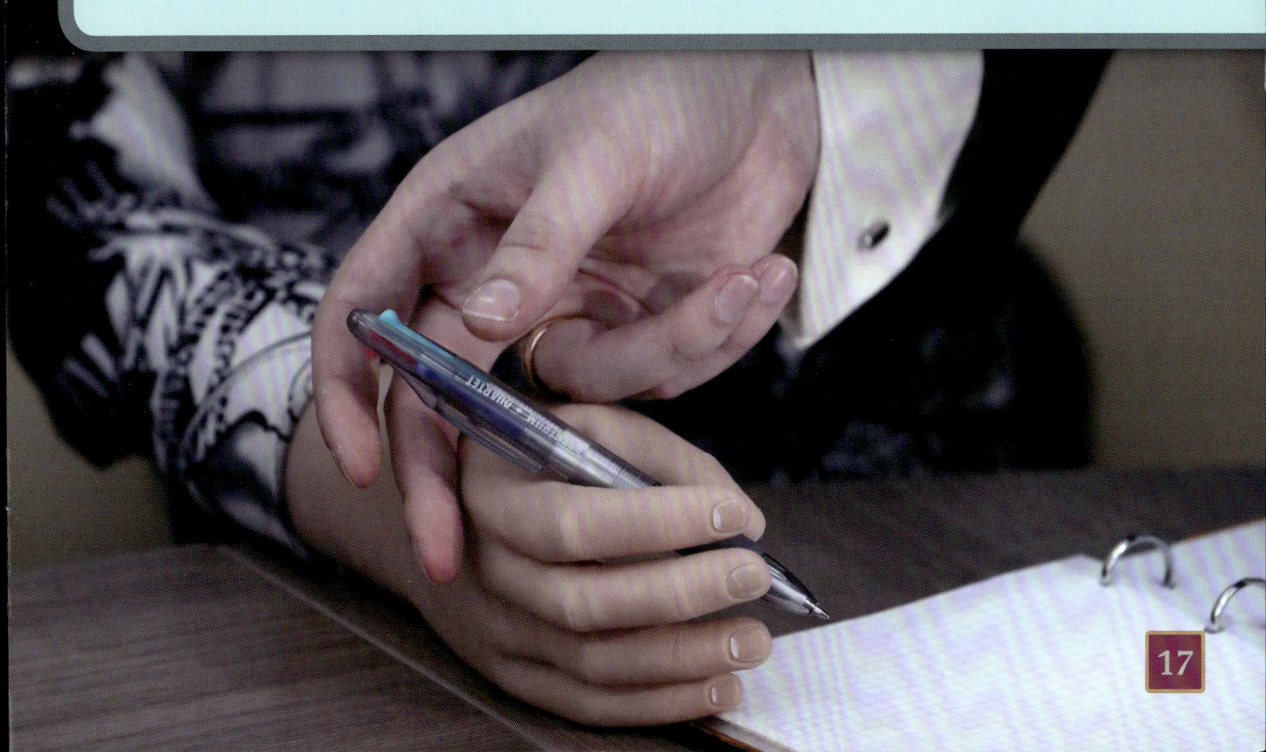

Lastly, advanced limb designs can be costly, and insurance companies often won't pay for them. Due to wear and tear, artificial limbs only last an average of three years. Although there are less expensive models, even those do not come cheap. With prices starting around $5,000 per piece, costs can skyrocket over a lifetime. Fortunately, some libraries and schools use their 3-D printers to make basic models and then donate the limbs they print.

If expenses can be kept low and other problems are resolved, many patients could live as if they never had the disease or injury that led to them needing a replacement part. Scientists are working on developing artificial versions of almost every organ in the body. In the future, we may be able to easily replace our worn-out or damaged parts. It sounds like science fiction, but it may be closer to reality than you think!

MORE TO EXPLORE

Cyborg is a superhero in the DC Comics universe. His many prostheses provide him with **SUPERHUMAN** strength, speed, and the power to fly.

Pony
Angel Marie, a pony whose front legs were accidently crushed by her mother when she was birthing her, now walks the pastures normally, thanks to a brace and a prosthetic limb.

Dolphin
When a two-month-old dolphin was found entangled in a fishing trap, veterinarians were able to save her life, but not her tail fluke. She flourished with a man-made tail and even appeared in her own Hollywood movie, *Dolphin Tale*.

Dog
A special therapy dog in Phoenix with four artificial paws—Chi Chi the golden retriever—brings smiles and comfort everywhere she goes, especially to fellow amputees.

Cat
Vincent, born without his back legs, was rescued from a shelter and given a new life. With surgery to insert prosthetic implants into his bones, this super cat now gets around with titanium legs.

Prostheses Are for Animals Too!

People aren't the only creatures to benefit from advancements in man-made body parts. Check out these cool creations that have helped a bevy of critters live and thrive.

Elephant
Pachyderm Chhouk was discovered in the Cambodian forest without his front left foot. Thanks to a series of prosthetics, walking is a cinch!

Tortoise
Fred lost her protective shell after it was damaged in a bushfire. Exposed, she was starving and sick, until her rescuers stepped in and got her a new 3-D printed one.

Toucan
In Costa Rica, a mutilated toucan was fitted with a nylon prosthetic beak, allowing him to eat on his own again and sing to his heart's content.

GLOSSARY

bypass
to go around something to avoid it

cardiac
related to the heart

chronic
occurring or continuing regularly

commandeered
taken over by force

digit
toe or finger

organ
a body part (such as the heart or kidney) that has a specific job

prostheses
man-made devices that can be used in place of an injured or missing body part

rejecting
when the body does not accept a new transplanted or artificial body part

stigma
a collection of negative beliefs that people have about something

trials
tests to determine whether a concept or idea works

viable
usable or workable

INDEX

3-D printing 14–15, 18
additive manufacturing 14
athletes 13
blood vessels 4, 15
bones 14
bypass 7
cardiac 9
cardiology 9
chronic 5
commandeered 16
computers 13, 16
costs 18
Cyborg 19
Dark Ages 10
dialysis 5
digit 3
donor 4, 6
Dr. Barney Clark 8
Dr. William H. Fissell 6
drawbacks 16
George Washington 2
hacking 16
heart 6, 7–9, 20
heart-lung machine 7–8
Jarvik 7 8
kidney 4–6, 20
limbs 10–13, 16–18
living tissue 15
lungs 7
organ 5–6, 7, 15, 17, 19
pacemaker 9
prostheses 3, 11–13, 16–17, 19
rejecting 6
stigma 13
teeth 2, 3, 14
The Kidney Project 6
total artificial heart (TAH) 8, 16
trials 15
viable 8
war 11–12